YOU WILL PASS ORGANIC CHEMISTRY

You Will Pass Organic Chemistry

Poetry Affirmations for Chemistry Students

Walter the Educator™

SKB

Silent King Books a WhichHead Imprint

"Earning a degree in chemistry changed my life!" - Walter the Educator. This book is dedicated to all the chemistry lovers, like myself, across the world.

CONTENTS

WHY I CREATED THIS BOOK?

Creating a poetry book to motivate students to pass the subject of Organic Chemistry is a unique and effective approach. Poetry has the power to engage emotions, capture attention, and simplify complex concepts. By presenting Organic Chemistry in a creative and relatable way, students may find it easier to grasp and retain the information. This book uses metaphors, analogies, and rhythmic verses to break down complex topics, making them more accessible and enjoyable. Including motivational poems that highlight the benefits of understanding Organic Chemistry can inspire students to put in the effort and succeed in the subject.

ONE

WONDERS OF ORGANIC CHEMISTRY

In the realm of atoms and bonds, a challenge lies,
Where molecules dance and electrons fly.
Organic Chemistry, a puzzle to solve,
A subject that tests the depths of resolve.

With formulas and reactions, it may seem tough,
But fear not, dear student, for knowledge enough.
Embrace the beauty of carbon's vast domain,
Unlock the secrets, let your passion sustain.

In the depths of the lab, where beakers collide,
A world of discovery, waiting inside.
From alkanes to aldehydes, let curiosity guide,
Every molecule a story, waiting to be untied.

Harness the power of functional groups,
Explore the pathways, where molecules stoop.

Understand the mechanisms, step by step,
Unveil the secrets, don't ever forget.
    Embrace the challenge, let failure not deter,
For perseverance, my friend, will help you conquer.
Seek guidance when needed, ask questions with might,
For knowledge is power, illuminating the night.
    Organic Chemistry, a subject of awe,
A journey of growth, a path to explore.
Let the bonds of determination be your guide,
And success in this subject, you shall surely abide.
    So, dear student, with passion and grit,
Conquer this subject, and never quit.
For in the realm of atoms and bonds, you'll see,
The wonders of Organic Chemistry, set yourself free.

TWO

DEAR STUDENT

In the realm of atoms, bonds entwined,
A subject vast, yet so refined.
Organic Chemistry, a world unseen,
Where molecules dance, and dreams convene.

From alkanes to ethers, reactions unfold,
A symphony of elements, stories untold.
Structures, mechanisms, and functional groups,
A puzzle to solve, like intricate loops.

Through the haze of equations, we shall tread,
With patience and passion, our fears shall shed.
For in every challenge lies hidden delight,
To conquer the mountain, we must take flight.

Aromatic compounds, aromatic bliss,
Unveiling the secrets, we cannot dismiss.
The beauty of carbon, its infinite forms,
A canvas of wonders, where knowledge transforms.

Embrace the mystery, let curiosity guide,
In the depths of this science, let your heart reside.
For with each discovery, a spark will ignite,
And the brilliance of learning will shine so bright.
Perseverance, dear student, is the key,
To unlock the doors and set yourself free.
In Organic Chemistry's intricate maze,
Lies the power to shape and amaze.
So fear not the challenges that lie ahead,
For with dedication, success shall be bred.
Passion and knowledge, hand in hand,
Together we'll conquer, and forever expand.

THREE

PERSEVERANCE AND DEDICATION

In the realm of atoms, bonds, and reactions,
Lies a subject that sparks both awe and passion.
Organic Chemistry, a wondrous domain,
Where mysteries and beauty forever remain.

Within its depths, molecules dance and sway,
A world of compounds, in a vibrant array.
From alkanes to esters, a symphony of creation,
Each structure a puzzle, ready for exploration.

Fear not, dear student, embrace this pursuit,
For Organic Chemistry holds treasures to loot.
Unlock the secrets, unravel the code,
And witness the wonders that it will unfold.

In labs and textbooks, knowledge shall thrive,
Reactions and mechanisms, you shall derive.

From synthesis to analysis, every step you take,
A path towards understanding, a bond you'll make.
So study with fervor, let curiosity guide,
For Organic Chemistry shall be your pride.
In every challenge, find strength and might,
And conquer the subject with all your might.
As you delve deeper, you'll witness the sublime,
The elegance of molecules, in perfect rhyme.
Organic Chemistry, a world to explore,
A journey of discovery forevermore.
With perseverance and dedication, you'll succeed,
Passing this subject, fulfilling your need.
So let the wonders of Organic Chemistry inspire,
And watch your knowledge and dreams transpire.

FOUR

YOU'LL FIND
YOUR WAY

In the realm of atoms, so small and unseen,
Lies a world of wonders, Organic Chemistry.
With bonds and reactions, a symphony of elements,
A puzzle to solve, intricate and relevant.
Through the haze of structures, complex and grand,
The student embarks on a journey, hand in hand.
With curiosity as their compass, they explore,
The mysteries of molecules, to their very core.
In the dance of electrons, they find their way,
Creating compounds, like art on display.
From alkanes to esters, each molecule unique,
A tapestry of life, waiting to speak.
Oh, student, fear not the challenges ahead,
For with each hurdle, you'll grow and spread.

Persevere with passion, let knowledge be your guide,
And soon you'll conquer this subject with pride.
    Organic Chemistry, a key to understand,
The intricacies of life, both near and far.
Unlock the secrets, unravel the code,
And in this pursuit, success shall unfold.
    So, my dear student, with determination and grace,
Embrace the beauty of Organic Chemistry's embrace.
For in this world of atoms, you'll find your way,
And pass this subject, triumphantly, one day.

FIVE

EMBRACE THE COMPLEXITIES

In the realm of atoms, where secrets hide,
Lies the world of Organic Chemistry, wide.
Curiosity, my dear student, is your guide,
To unlock the wonders that within reside.
Let passion ignite your eager mind,
For in carbon's playground, treasures you'll find.
Bonds and reactions, a symphony of art,
Unraveling molecules, a dance with your heart.
Perseverance, my dear, is the key,
Through formulas and mechanisms, you'll see.
Each failure a lesson, each challenge a test,
But with steadfast resolve, you'll conquer the rest.
Aromatic compounds, fragrant and bold,
Their structures a story waiting to be told.

From benzene rings, a world unfurls,
Where atoms align like precious pearls.
    Seek knowledge, my student, with unyielding zest,
For Organic Chemistry will put you to the test.
Let the beauty of molecules light up your way,
As you unravel their secrets, day by day.
    With every bond formed, every reaction complete,
You'll find strength in your journey, never retreat.
Embrace the complexities, let them inspire,
And watch your understanding reach higher.
    Organic Chemistry holds treasures untold,
For those who dare to be brave and bold.
With dedication and curiosity, you'll succeed,
Unlocking the wonders you've always believed.
    So, my dear student, let this poem be your voice,
As you embark on this journey, make your choice.
In passing this subject, your dreams will take flight,
Let Organic Chemistry guide you, with all its might.

SIX

NO FEARS

In the realm of atoms, a world unseen,
Lies the mystic land of Organic Chemistry.
A subject of wonder, a puzzle to solve,
Where bonds and molecules beautifully evolve.
    Embark on this journey, young scholar of mine,
Unlock the secrets, let your passion shine.
For in the depths of this scientific art,
Lies a realm of knowledge waiting to impart.
    From alkanes to alkenes, functional groups abound,
Reactions and mechanisms, let your mind astound.
With every step you take, a discovery awaits,
A glimpse into nature's intricate traits.
    Through perseverance and unwavering will,
You'll conquer the challenges, your dreams fulfill.

For Organic Chemistry, though complex it may be,
Rewards the curious with answers to see.
    Embrace the structure, the bonds that unite,
Compounds and compounds, a symphony of might.
Seek understanding, let curiosity guide,
And in this subject's depths, you'll surely abide.
    So fear not the formulas, the names that confound,
For in this subject's playground, brilliance is found.
With passion and dedication, you shall surpass,
And Organic Chemistry's mysteries amass.
    Believe in yourself, let knowledge be your key,
Unlock the wonders of Organic Chemistry.
For in your hands, lies the power to transform,
To shape the world with every lesson learned.
    So study with purpose, let your dreams ignite,
And Organic Chemistry's challenges, you shall smite.
For success awaits the student who perseveres,
In this remarkable subject, with no fears.

SEVEN

STRENGTH TO SURPASS ANY ODDS

In the realm of atoms, where mysteries reside,
Lies a subject called Organic Chemistry, wide.
With molecules dancing, in elegant array,
It beckons you closer, to learn and to play.
    Organic Chemistry, a puzzle to unfold,
A world of reactions, both precious and bold.
From alkanes to aldehydes, each bond holds a key,
Unlocking the secrets of this grand chemistry.
    Amidst the formulas and structures, so complex,
Lies the power to learn, to excel, to impress.
A student, like you, with passion untamed,
Can conquer this subject, unafraid and unchained.
    Let the beauty of carbon ignite your desire,
To delve into molecules, with passion on fire.

For in every atom, a story unfolds,
Of life's building blocks, in patterns untold.
Persevere through the challenges, embrace the unknown,
With each step you take, your knowledge will have grown.
Curiosity shall be your guiding star,
As you navigate the wonders, both near and far.
Organic Chemistry, a journey of the mind,
Unveiling the secrets that nature designed.
So study with purpose, with dedication so true,
And Organic Chemistry shall lead you to breakthrough.
In this intricate dance of atoms and bonds,
You'll find the strength to surpass any odds.
Embrace the challenges, let knowledge be your guide,
And Organic Chemistry will forever be your pride.

EIGHT

VICTORY IN ORGANIC CHEMISTRY

In the realm of atoms and bonds,
Where Organic Chemistry absconds,
A world of wonders awaits your gaze,
Embrace the challenges that set ablaze.
    From alkanes to aldehydes,
Molecules dance and collide,
With carbon's versatile embrace,
A symphony of life takes place.
    In the depths of aromatic rings,
Aromaticity's melody sings,
Resonance cascades through the air,
Unleashing knowledge, beyond compare.
    Like a puzzle, intricate and grand,
Molecular structures stand,

Connect the dots, unlock the key,
And marvel at the complexity.
  Let reactions guide your way,
As synthesis paints a colorful display,
From esters to ethers, a vibrant blend,
In Organic Chemistry, success will ascend.
  Though mechanisms may confound,
With perseverance, they can be unwound,
For every challenge that you face,
Unveils a deeper level of grace.
  So, fear not the exams that loom,
For within you, knowledge will bloom,
Embrace the beauty, the art, the strife,
And conquer Organic Chemistry with life.
  For in this subject lies a treasure trove,
Where passion and dedication rove,
Unlock the secrets, let understanding soar,
And victory in Organic Chemistry is yours to explore.

NINE

YOU'LL LEARN AND GROW

In the realm of molecules and bonds,
Where atoms dance and connections spawn,
Lies the world of Organic Chemistry,
A subject of beauty, complexity.
    Fear not the formulas and reactions,
For within them lie hidden attractions.
Unlock the secrets of carbon's art,
And you'll find a path straight to the heart.
    From alkanes to aldehydes, a vast array,
Each compound has stories to convey.
Embrace the challenge with open arms,
And witness the wonders, like nature's charms.
    Through esters and ethers, journey deep,
Explore the realm where compounds sleep.

With every challenge, you'll learn and grow,
Unraveling mysteries, high and low.
	In Organic Chemistry, dreams take flight,
As you delve into the world of light.
From synthesis to mechanisms unknown,
You'll discover a world uniquely shown.
	So, let perseverance be your guide,
With every failure, don't step aside.
For in the struggle, you'll find your way,
To success and triumph, come what may.
	Believe in yourself, let passion ignite,
As you conquer Organic Chemistry's might.
With knowledge and determination in your core,
Pass the subject and unlock boundless doors.

TEN

PASSION AND DEDICATION

In the realm of atoms and bonds,
Where beauty and complexity respond,
There lies the subject, Organic Chemistry,
A puzzle to solve, a mystery.

In molecules, a symphony unfolds,
Elements dance, their tale untold,
Carbon, hydrogen, oxygen in harmony,
Creating life's intricate tapestry.

Embrace the challenges, dear student,
For they shape you, make you resilient,
In the lab, where experiments thrive,
Unlock the secrets, let your passion drive.

From alkanes to aldehydes, reactions galore,
Each step a triumph to explore,

Aromatic compounds, aromatic bliss,
A world of possibilities, don't dismiss.
    The periodic table, a map of creation,
Elements in harmony, a cosmic revelation,
Organic Chemistry, the language of life,
Unveiling the secrets, dispelling the strife.
    Let the molecules guide you, my friend,
Through the bonds that intertwine and blend,
For within their intricate design,
Lies the power to shape and redefine.
    So fear not the formulas, the structures unknown,
Organic Chemistry, a garden to be sown,
With patience and perseverance, you shall prevail,
Unlocking the wonders, a triumphant tale.
    For in the realm of atoms and bonds,
A world of discovery awaits beyond,
Passion and dedication, your guiding light,
May you conquer the subject with all your might.

ELEVEN

BELIEVE IN YOURSELF

In the realm of atoms, where secrets lie,
Organic Chemistry's beauty does imply.
A dance of molecules, elegant and grand,
A subject that few can fully understand.
  Through bonds and reactions, a world unfolds,
A tapestry of elements, stories untold.
From hydrocarbons to alcohols so pure,
Organic Chemistry beckons, an adventure sure.
  In each compound's structure, a puzzle to solve,
A challenge to conquer, to evolve.
With patience and diligence, you'll find the way,
To pass this subject, and seize the day.
  Embrace the complexity, don't be afraid,
For in the depths of knowledge, wisdom is laid.

The periodic table, a guidebook to wield,
Unlocking the secrets, a treasure revealed.
    With each equation, a step closer you'll be,
To mastering reactions, so wild and free.
Synthesis and mechanisms, they hold the key,
To understanding the wonders of chemistry.
    So let your curiosity ignite,
As you dive into the world, shining so bright.
With perseverance and passion, you'll surely excel,
Organic Chemistry's triumph, your story to tell.
    Believe in yourself, and don't ever doubt,
You have the power, there's no need to pout.
For in your hands lies the chance to surpass,
Organic Chemistry's challenges, with flying colors,
amass.

TWELVE

BRILLIANT FLASH

In the realm of atoms and bonds we delve,
Where Organic Chemistry casts its spell,
A subject so vast, a mystery to unveil,
Let perseverance guide you, and never fail.

Within the molecules, a world takes form,
Unraveling secrets, the beauty will transform,
Like an alchemist, you hold the key,
To unlock the wonders that lie before thee.

In carbon's dance, compounds arise,
Reactions ignite like fireflies,
Embrace the challenges, embrace the strife,
For in the struggle lies the essence of life.

From hydrocarbons to functional groups,
A symphony of elements, harmonious loops,

Curiosity is the compass that will lead,
To a realm of knowledge, where you'll succeed.
    The bonds may break, the molecules shift,
But your passion and dedication will uplift,
With every failure, a lesson to learn,
A stepping stone towards success you'll earn.
    Organic Chemistry, a puzzle to decipher,
A journey of discovery that will never tire,
So fear not the complexity, fear not the test,
For with determination, you'll be the best.
    Believe in yourself, embrace the unknown,
The world of Organic Chemistry you'll own,
With passion and knowledge, you'll surpass,
And conquer this subject like a brilliant flash.

THIRTEEN

SUCCESS IS FOR YOU

In the realm of atoms, where wonders reside,
Lies the subject of Organic Chemistry, with secrets untied.
A symphony of molecules, dancing in harmony,
Unveiling the beauty of life's grand tapestry.

Carbon, the backbone, the foundation of all,
Building blocks of existence, both big and small.
From hydrocarbons to functional groups,
A universe within, where knowledge loops.

Bonds interweaving, creating endless chains,
A puzzle to solve, where brilliance remains.
Resonance and isomerism, puzzles to be solved,
An intellectual odyssey, where dreams evolve.

Fear not the mechanisms, the reactions so vast,
For within your heart, the fire burns fast.

Embrace the challenge, with passion so pure,
Organic Chemistry's allure, you shall endure.
The laboratory, your sanctuary of creation,
Where experiments unfold, in pure fascination.
With flask in hand and goggles on your eyes,
You'll unravel the mysteries, reaching for the skies.
Through dedication and perseverance, you shall prevail,
Unlocking the secrets, that others may fail.
So let your curiosity guide your way,
As you embark on this journey, day by day.
Organic Chemistry, a subject profound,
A key to understanding, the world all around.
Let not the equations and structures intimidate,
For with knowledge and courage, you shall dominate.
Passion and persistence, your allies true,
In the realm of Organic Chemistry, success is for you.

FOURTEEN

LET YOUR DREAMS TAKE SHAPE

In the realm of atoms, where secrets reside,
Lies a world of wonders, Organic Chemistry's tide.
A symphony of molecules, a dance of bonds,
Unlocking the universe, where knowledge responds.
Let curiosity be your guiding light,
Embrace the challenges, with all your might.
For in the depths of synthesis, a masterpiece awaits,
A canvas of compounds, where creation resonates.
Mechanisms unravel, like delicate threads,
Revealing the mysteries that nature spreads.
The puzzle of reactions, a puzzle to unfold,
With perseverance and passion, let your story be told.
The periodic table, a map to explore,
Elements and compounds, a treasure trove in store.

From carbon's embrace to oxygen's breath,
Discover the language of atoms, a journey with no death.
    Believe in yourself, for you possess the key,
To tame the complexities, to set your knowledge free.
With every step you take, the path becomes clearer,
Organic Chemistry, a subject to revere.
    So fear not the challenges, let passion ignite,
For in the depths of Organic Chemistry, you'll take flight.
Unlock the secrets, let your dreams take shape,
And conquer this subject, with knowledge as your cape.

FIFTEEN

YOU HOLD THE POWER

In the realm of atoms, bonds intertwine,
Organic Chemistry, a puzzle divine.
With carbon's dance, and molecules so vast,
A world of wonders to uncover, unsurpassed.
Let passion ignite like a fiery flame,
Embrace the complexity, don't shy away from the game.
For in this realm of atoms, secrets reside,
Unveiling the mysteries, with knowledge as our guide.
Curiosity, the key to unlock the gate,
To the beauty and wonders that lie in wait.
Let it drive you forward, like a gust of wind,
Through the valleys and peaks, where discoveries begin.

Perseverance, oh steadfast friend,
In the face of challenges, it will transcend.
For every obstacle, a lesson to be learned,
In the depths of struggle, your strength will be earned.
Believe in yourself, for you hold the power,
To conquer the subject, in every hour.
With determination and focus, you'll go far,
Unleashing your potential, like a shining star.
Organic Chemistry, a world to explore,
A subject that opens boundless doors.
Embrace its complexities, let your passion soar,
And success will be yours, forevermore.

SIXTEEN

GREATNESS AWAITS

In the realm of atoms and bonds so fine,
Where elements dance and molecules align,
Lies the beauty of Organic Chemistry,
A world of wonders, waiting for you to see.

Unlock the secrets of carbon's embrace,
And witness nature's elegant grace,
From hydrocarbons to functional groups,
A symphony of reactions, like magic hoops.

With beakers and flasks, your tools in hand,
Embark on a journey, a dream so grand,
The path may be winding, the challenges tough,
But fear not, dear student, for knowledge is enough.

Organic Chemistry, a gateway to explore,
The mysteries of life, to unravel and adore,
From DNA strands to drugs that heal,
Your understanding, a power to reveal.

So embrace the bonds, both weak and strong,
Let curiosity guide you along,
With perseverance as your faithful guide,
Success shall be yours, let doubt subside.
For in this subject, greatness awaits,
A world of discoveries, for you to create,
Believe in yourself, let passion ignite,
And Organic Chemistry shall be your guiding light.

SEVENTEEN

EXPLORE, LEARN, AND PROVE

In the realm of atoms, bonds, and reactions,
Lies a subject that sparks both awe and dissatisfac-
tion,
Organic Chemistry, a puzzle to unravel,
A journey that requires courage to travel.

Fear not the complex structures, my dear student,
For within lies the path to knowledge, resplendent.
Like a painter with a palette of elements,
You have the power to create wonders, evident.

From alkanes to aldehydes, and esters so sweet,
The world of molecules lays at your feet.
Through resonance and aromaticity's embrace,
Unlock the secrets of this captivating maze.

Though the road may be arduous and long,

Let perseverance and determination be strong.
For with each challenge you overcome,
Your understanding of the subject will blossom.
    Embrace the beauty of carbon's dance,
And watch your passion for chemistry enhance.
For in conquering this subject, you'll find,
A sense of accomplishment, one of a kind.
    So, dear student, with your heart ablaze,
Dive into the depths of Organic Chemistry's maze.
For within its intricacies lies a treasure trove,
Waiting for you to explore, learn, and prove.

EIGHTEEN

OVERCOME ANY STRESS

In the realm of atoms and bonds,
Where molecules dance and respond,
Lies a world of wonder and intrigue,
Where Organic Chemistry takes the lead.
    Embrace this subject with all your might,
Unveil its secrets, unlock its light,
For in its depths, beauty resides,
A symphony of elements collides.
    From alkanes to alcohols, it unfolds,
A puzzle of structures to behold,
With every reaction, a new creation,
A chemical journey of revelation.
    Let passion guide you on this path,
Through formulas and compounds, you'll laugh,

For in each carbon, hydrogen, and oxygen,
Lies the power to transform and envision.
   Yes, there may be challenges along the way,
But fear not, for knowledge will hold sway,
With perseverance and dedication, you'll find,
The strength to conquer, the wisdom to shine.
   So, brave student, let your spirit soar,
Through the wonders of Organic Chemistry's lore,
With curiosity as your constant guide,
Success in this subject, you shall ride.
   Believe in yourself, for you possess,
The courage to overcome any stress,
And as you pass through this academic test,
A world of possibilities, you'll manifest.

NINETEEN

UNLOCK THE SECRETS

In the realm of atoms, where bonds are formed,
Lies a subject that leaves many minds warmed.
Organic Chemistry, a captivating art,
Unfolding the secrets of nature's own heart.
    From carbon's dance to molecular symphony,
A world of compounds awaits your discovery.
Fear not the reactions, the mechanisms unknown,
For in every challenge, true knowledge is sown.
    In the vast ocean of compounds, you'll find,
An endless array, a dazzling design.
Each molecule, a story waiting to be told,
A puzzle to solve, a mystery to unfold.
    With patience and focus, you'll navigate,
Through resonance structures and isomers so great.

Synthesis and mechanisms, a journey of delight,
With every step forward, a new insight.
Embrace the complexities, the beauty within,
For in Organic Chemistry, wonders begin.
Let your passion guide you, like a guiding star,
And success will be yours, no matter how far.
Believe in yourself, for you hold the key,
To unlock the secrets, in this chemistry.
With dedication and perseverance, you'll soar,
And Organic Chemistry, you'll truly adore.
So, dear student, take heart and don't despair,
Organic Chemistry's challenges, you'll dare.
For in conquering this subject, you'll find,
A world of knowledge and wonders unconfined.

TWENTY

VIVID REALIZATION

In the world of atoms and bonds we delve,
Where molecules dance and secrets dwell.
Organic Chemistry, a wondrous realm,
Where miracles of life truly overwhelm.
    Fear not the equations, complex and long,
For within their depths, knowledge grows strong.
A symphony of reactions, a delicate chore,
Unveiling nature's mysteries, forevermore.
    From alkanes to aldehydes, each compound unique,
Unlocking the beauty, is what we seek.
With every bond broken, and every bond made,
A world of possibilities, we shall create.
    Like a painter with colors, we mix and blend,
Creating molecules, our art will transcend.

The power within you, a genius untapped,
Embrace the challenge, and success will unwrap.
    Though the journey may be filled with strife,
Persevere, dear student, and embrace this life.
For in the realm of Organic Chemistry,
Lies the key to unlock your true destiny.
    Believe in yourself, let passion ignite,
With determination, you'll conquer the fight.
So study with fervor, with heart and dedication,
And watch as your dreams become a vivid realization.

TWENTY-ONE

TRANSFORMATION OF YOUR ASPIRATIONS

In the realm of atoms and bonds,
Where knowledge and curiosity respond,
Lies the enigma of Organic Chemistry,
A subject that tests your perseverance and ability.
    Fear not the complexity of molecules,
For within them lies beauty that transcends all rules.
Unlock the secrets they hold within,
And let your passion for learning begin.
    With every reaction and mechanism,
Comes an opportunity for wisdom.
Through resonance and electron flow,
You'll witness the magic that molecules bestow.
    Embrace the challenges that lie ahead,
For they will shape you, like the molecules you tread.

Let your mind soar through the realm of compounds,
And watch as your understanding compounds.
 Believe in yourself, with unwavering faith,
For knowledge is the key that unlocks the gate.
Persevere through the reactions and equations,
And witness the transformation of your aspirations.
 Organic Chemistry, a puzzle to solve,
A journey of discovery that will evolve.
With dedication and a thirst to learn,
Success in this subject, you will discern.
 So, dear student, take up the noble quest,
And let Organic Chemistry put you to the test.
For within its intricacies, you'll find,
The power to expand your brilliant mind.

TWENTY-TWO

KNOWLEDGE SETS
YOU FREE

In the realm of atoms, where secrets reside,
Lies a subject called Organic Chemistry, a wondrous ride.
With compounds and reactions, it beckons you near,
To delve into its depths, without any fear.
The bonds that unite, so intricate and strong,
Forming molecules, where beauty belongs.
From hydrocarbons to functional groups,
A world of possibilities, ready to be scooped.
Oh, student of science, hear my plea,
Embrace this subject, set your knowledge free.
For in Organic Chemistry, lies a key,
To understanding life's complexity.
Through labs and textbooks, you'll stumble and fall,
But don't lose heart, for you'll conquer it all.

With perseverance and passion, you'll rise above,
Unlocking the mysteries, fueled by your love.
    Believe in yourself, for you hold the power,
To grasp the concepts, in this chemical tower.
With every equation and mechanism you learn,
A brighter future, you'll surely earn.
    So study with fervor, and don't let go,
Organic Chemistry's wonders, you'll come to know.
For within its challenges, lies hidden treasure,
A world of discovery, beyond all measure.
    Pass this subject, and you'll find the key,
To a future where knowledge sets you free.
Embrace the beauty, let it ignite your flame,
And Organic Chemistry shall forever acclaim.

TWENTY-THREE

PPD

In the realm of atoms and bonds we delve,
Where Organic Chemistry's secrets dwell.
A subject demanding focus and grace,
A challenge to conquer, at our own pace.

Through molecules and reactions we explore,
A symphony of elements we adore.
With every carbon, hydrogen, and bond,
A world of possibilities we respond.

Let passion fuel your curious fire,
Ignite the desire to reach higher.
For in the realm of Organic Chemistry,
Lies the gateway to scientific victory.

Persevere through mechanisms complex,
Unravel the puzzles, never perplexed.
With every failure and every try,
The path to success steadily draws nigh.

Dedicate yourself to the noble cause,
Unleash your potential, shatter all flaws.
For Organic Chemistry holds the key,
To unlock the brilliance that lies within thee.
Embrace the challenges that lie ahead,
With determination, no fear to dread.
Believe in yourself, for you have the power,
To conquer this subject, hour by hour.
So, rise up, student, and face the test,
Organic Chemistry, you shall attest.
With ppd (passion, perseverance, and dedication),
You'll pass with pride, a celebration.

TWENTY-FOUR

FASCINATING CLASS

In the realm of atoms, bonds, and compounds,
Lies a world of wonders, where knowledge abounds.
Organic Chemistry, a puzzle to unravel,
A subject that challenges, yet promises to marvel.
   Fear not the formulas, the reactions, and reagents,
For within lies the beauty, the secrets it presents.
From alkanes to alcohols, esters to ethers,
A symphony of molecules, dancing together.
   Embrace the challenges, let curiosity ignite,
For in the depths of Organic Chemistry, lies your true might.
With each carbon chain, a story unfolds,
A chance to explore, to create, to behold.
   Persevere through the obstacles, the doubts that arise,

For within you, a flame of potential lies.
Believe in yourself, let passion be your guide,
And Organic Chemistry shall be your stride.
Unlock the mysteries, unravel the unknown,
Expand your horizons, let your knowledge be sown.
For in this subject, lies a world of possibilities,
A gateway to innovation, to scientific discoveries.
So, student, embrace this journey with zest,
And Organic Chemistry will put you to the test.
With dedication and belief, you shall surpass,
For success awaits, within this fascinating class.

TWENTY-FIVE

YOUR TRUE POTENTIAL

In the realm of atoms, bonds, and reactions,
Lies a world of wonder, Organic Chemistry's attractions.
A subject that may seem daunting at first,
But fear not, dear student, quench that thirst.
　　For in the depths of this mysterious domain,
Lies the key to understanding life's intricate chain.
From the simplest compounds to complex creations,
Organic Chemistry unveils nature's revelations.
　　Each element, each molecule, holds a story untold,
Waiting to be discovered, waiting to unfold.
Unlock the secrets, delve deep into the unknown,
And watch as your knowledge and confidence have grown.

Yes, it may be challenging, a true test of will,
But remember, with perseverance, you can fulfill.
Embrace the journey, embrace the strife,
For within these trials lies the beauty of life.
So study with passion, let curiosity guide your way,
And watch as the pieces fall into place, day by day.
Organic Chemistry is a gateway to the unknown,
A path that leads to wisdom, a path that's your own.
Believe in yourself, dear student, you have what it takes,
To conquer this subject, whatever it may take.
Embrace the challenge, embrace the thrill,
And witness the power of your determined will.
Organic Chemistry, a puzzle to unravel,
But with dedication, success you shall travel.
So rise above the doubts, spread your wings and fly,
For in this subject, your true potential lies.

TWENTY-SIX

TESTAMENT TO YOUR STRENGTH

In the realm of atoms, bonds, and reactions,
Lies a subject that sparks both awe and satisfaction.
Organic Chemistry, a puzzle to unravel,
A journey that will test, but never unravel.
    Oh, student of science, fear not this path,
For within its depths, lies infinite worth.
Unlock the secrets of carbon's dance,
And witness the miracles of circumstance.
    From alkanes to aldehydes, esters to amines,
Organic compounds, a symphony divine.
Embrace each equation, each structure and name,
For they hold the key to nature's mysterious game.
    Let curiosity be your guiding light,
As you venture into this world, shining bright.

With each experiment and lab report,
You'll uncover knowledge, you'll ignite a spark.
Through mechanisms and reactions, you'll find,
That Organic Chemistry is truly one of a kind.
It challenges your mind, expands your view,
And reveals the wonders that science can do.
So, heed my words, dear student, with grace,
Embrace the challenge, run this noble race.
For in the realm of Organic Chemistry,
Lies the power to transform, to set your mind free.
Passion and perseverance shall be your guide,
As you conquer this subject, side by side.
Believe in yourself, let your potential bloom,
And witness the magic in this chemical room.
Organic Chemistry, a gateway to knowledge,
A testament to your strength, your courage.
So, rise above the doubts, embrace the sublime,
For success in this subject shall be truly thine.

TWENTY-SEVEN

RESONANCE AND REACTIONS

In the realm of atoms, bonds intertwine,
Organic Chemistry, a puzzle divine.
A subject so vast, yet captivatingly fine,
Let your passion for it forever shine.

Embrace the molecules, their structures so grand,
Unlock the secrets, let your mind expand.
Through resonance and reactions, you will understand,
The beauty of nature, at your command.

From alkanes to aldehydes, each compound unique,
A symphony of elements, a language you'll speak.
With every reaction, progress you'll seek,
Transforming matter, the future you'll tweak.

Yes, the road may be tough, the challenges steep,

But fear not, dear student, for knowledge runs deep.
With each equation, a puzzle to keep,
Solve them with diligence, and success you'll reap.
    For Organic Chemistry, a gateway to explore,
A world of possibilities, waiting to adore.
Believe in yourself, let your passion soar,
And the wonders of science, you'll forever adore.
    So study with dedication, let your dreams take flight,
Organic Chemistry, a beacon of light.
In its depths, lies knowledge and might,
Unleash your potential, embrace the delight.

TWENTY-EIGHT

JOURNEY OF DISCOVERY

In the realm of atoms, where molecules collide,
Lies the subject of Organic Chemistry, deep and wide.
A world of carbon chains and functional groups,
Where bonds are formed and reactions take their loops.
Oh, student of science, with a heart full of fire,
Embrace the challenge, let your passion inspire.
For within these structures, secrets are unveiled,
Unlocking the mysteries, where knowledge is hailed.
With perseverance as your guiding light,
Navigate the pathways, conquer the night.
Through resonance and mechanisms, you shall prevail,
And witness the beauty that these compounds entail.
From alkanes to aldehydes, esters to amines,

Each compound a puzzle, waiting to align.
With every equation and synthesis you undertake,
You'll find the satisfaction no other subject can make.
    So, march forward, student, with determination and might,
Let your curiosity be your guiding light.
For in the realm of Organic Chemistry, you'll see,
A world of possibilities, waiting for you to set free.
    Embrace the challenge, let your passion ignite,
Organic Chemistry, a subject worth the fight.
Believe in yourself, for you have what it takes,
To conquer this subject, and celebrate your ache.
    So, study hard, with dedication and zeal,
And watch as Organic Chemistry reveals,
The wonders it holds, the knowledge it imparts,
A journey of discovery, where you'll find your heart.

TWENTY-NINE

PASSIONATE STUDENT

In the realm of molecules, where secrets lie,
There blooms a subject, Organic Chemistry, high.
With atoms dancing, and bonds that intertwine,
A world of wonders, waiting to be defined.
   Oh, student, hear my humble plea,
Dive into this realm, and set your spirit free.
For in the depths of carbon's endless chains,
Lies the power to unlock knowledge's gains.
   Through Lewis structures and resonance forms,
You'll traverse a world where molecules perform.
Embrace the challenges, let them fuel your fire,
For with each obstacle, your strength will aspire.
   In lab coats and goggles, you'll find your way,
Exploring reactions, day after day.

From alkanes to alcohols, reactions abound,
Witness the magic as new compounds are found.
    And when doubt tries to cloud your mind,
Remember, dear student, you're one of a kind.
With perseverance and belief in your heart,
You'll conquer Organic Chemistry, playing your part.
    So study the mechanisms, memorize the names,
Let the beauty of molecules fuel your flames.
For in this subject lies a world unseen,
Where discoveries await, like dreams in a dream.
    Passionate student, don't shy away,
Embrace the challenges, come what may.
For Organic Chemistry holds the key,
To unlock the potential within you and me.

THIRTY

DON'T BE DISMAYED

In the realm of atoms and bonds, you'll find,
A subject that may seem complex, but don't mind,
Organic Chemistry, a world so vast,
Where wonders and possibilities are amassed.

From molecules to reactions, it's a symphony,
A dance of electrons, a captivating melody,
Embrace the challenge, don't be dismayed,
For in this subject, your brilliance will be displayed.

With dedication and passion, you'll unravel the code,
Unlocking the secrets, as your knowledge unfolds,
Each equation and mechanism, a puzzle to solve,
As you delve deeper, your problem-solving evolves.

The beauty of Organic Chemistry lies,
In the patterns and structures that mesmerize,

A world of compounds, their properties and reactivity,
Hold the key to a future of boundless creativity.
Believe in yourself, for you have the power,
To conquer this subject, to bloom like a flower,
Let curiosity guide you, let it fuel your flame,
And watch as Organic Chemistry becomes your claim.
So, dear student, fear not the challenges ahead,
For in Organic Chemistry, your dreams will be fed,
Passion and persistence, your allies to be,
And success in this subject, you shall see.

THIRTY-ONE

WAITING FOR THEE

In the realm of atoms, where secrets lie,
Organic Chemistry's mysteries reside.
A subject complex, yet wondrous and grand,
Unveiling the beauty of molecules, hand in hand.
　　With carbon as its backbone, life it sustains,
Creating bonds, reactions that never wane.
From alkanes to aldehydes, a vast array,
Organic Chemistry beckons, come and play.
　　Through resonance and aromaticity,
Discover the magic of electron density.
Unleash your curiosity, let it soar,
In this realm of compounds, an endless lore.
　　Aromatic rings, their fragrance divine,
Unraveling reactions, a dance so fine.

Stereochemistry, chirality's grace,
Unlocking the secrets of three-dimensional space.
    Yes, there are hurdles to overcome,
But with each challenge, you'll surely become,
A master of mechanisms, reactions so fast,
Transforming molecules, a power unsurpassed.
    So fear not, dear student, embrace the fight,
For Organic Chemistry holds the light.
In its depths lies the key to your success,
Unleashing your potential, no less.
    With perseverance and passion, you shall prevail,
Conquer the subject, let your knowledge unveil.
For in this journey, you'll find your way,
Organic Chemistry, guiding you day by day.
    Believe in yourself, your abilities untold,
Organic Chemistry's wonders, behold!
You'll pass with flying colors, you'll see,
A world of possibilities, waiting for thee.

THIRTY-TWO

CONQUERING ORGANIC CHEMISTRY

In the realm of atoms, a world unfolds,
Where Organic Chemistry's tale is told.
A subject of wonders, complex and vast,
A challenge to conquer, it beckons you fast.

Fear not the formulas, the bonds that bind,
For within them, knowledge you shall find.
Like puzzle pieces, they fit in place,
Unlocking the secrets of carbon's grace.

From alkanes to aldehydes, a symphony of sorts,
Building blocks of life, where it all starts.
Functional groups, reactions, and more,
A universe of molecules to explore.

Through each difficulty, you'll learn and grow,
Persistence and passion, your strength will show.

The path may be winding, the road may be tough,
But with determination, you'll rise above.
    Believe in yourself, your abilities shine,
With hard work and focus, success will be thine.
Embrace the beauty of atoms in motion,
For in their dance lies the potion.
    Organic Chemistry, a subject so grand,
Unlock its secrets with your steady hand.
Let curiosity be your guiding light,
And watch the wonders unfold within your sight.
    So persevere, dear student, don't give in,
With every challenge, let your knowledge begin.
For in conquering Organic Chemistry's quest,
You'll find the rewards are truly the best.

THIRTY-THREE

DETERMINATION BE YOUR GUIDE

In the realm of atoms, where wonders reside,
Lies the realm of Organic Chemistry, wide.
It beckons you, student, to embark on this quest,
To unravel the secrets, to be your very best.

From carbon's dance to molecules so grand,
A symphony of elements, at your command.
Bonds and reactions, they intertwine,
Creating a tapestry, both complex and divine.

Organic Chemistry, a puzzle to solve,
With patience and diligence, problems evolve.
Each compound, a challenge, a riddle to crack,
But fear not, dear student, you'll stay on track.

Through resonance structures and functional groups,
groups,

You'll navigate the maze, dispelling the loops.
The beauty of molecules, their intricate design,
Will captivate your mind, like stars that align.
      Embrace the mechanisms, the pathways they trace,
For within lies the key to unlock every space.
The power to create, to synthesize new,
Organic Chemistry unveils a world for you.
      So, let your passion ignite, like flames in the night,
For Organic Chemistry holds endless delight.
Believe in yourself, let determination be your guide,
And soon, dear student, you'll pass with pride.
      For within these atoms, a universe to explore,
Where knowledge awaits, like never before.
Unlock the doors of Organic Chemistry's domain,
And let your success be your sweetest refrain.

THIRTY-FOUR

TRIUMPHANT TALE

In the realm of molecules, where wonders reside,
Lies Organic Chemistry, a captivating guide.
A subject complex, yet filled with allure,
Where atoms dance, and bonds endure.

Fear not, dear student, for I shall declare,
The beauty that lies in this scientific affair.
Organic compounds, with structures so grand,
Hold the key to a world within your hand.

From alkanes to aldehydes, ketones to esters,
Each molecule a puzzle, a quest to master.
Through resonance and reactions, a pathway unfolds,
Leading to discoveries, more precious than gold.

Let curiosity be your faithful guide,
As you explore the secrets, nowhere to hide.

Embrace the challenges, with a heart so bold,
For in the depths of knowledge, your story unfolds.
Believe in yourself, your potential untold,
For Organic Chemistry is where dreams unfold.
With passion and perseverance, you shall prevail,
Unlocking the mysteries, like a triumphant tale.
So, dear student, let your spirit ignite,
For Organic Chemistry holds a future so bright.
With dedication and hard work, success will be near,
And the rewards of your efforts, oh so clear.
In this world of atoms, where possibilities lie,
Organic Chemistry will be your ally.
So, march forth with courage, embrace the unknown,
For in conquering this subject, true strength is grown.

THIRTY-FIVE

A JOURNEY DIVINE

In the realm of atoms and bonds, behold,
A subject of wonders, yet untold.
Organic Chemistry, a realm of might,
Where elements dance in captivating light.

From carbon's embrace to molecules' delight,
A symphony of reactions, day and night.
The secrets of life lie within its core,
A gateway to wisdom, forevermore.

Fear not the complexity, young soul,
For Organic Chemistry has stories yet untold.
Embrace the challenge, let passion ignite,
As you journey through each chemical rite.

Through Lewis structures and resonance's glow,
Discover the patterns, let your knowledge grow.

Unlock the mysteries of functional groups,
And witness the power of molecular loops.
    In labs and lectures, let your mind roam,
Unleash your potential, let curiosity bloom.
With determination as your guiding star,
You'll conquer each hurdle and venture far.
    For Organic Chemistry is a realm of might,
A canvas to paint dreams in colors bright.
So, believe in yourself, let your spirit rise,
And watch as success materializes.
    In each reaction, find joy and delight,
As you unravel the beauty, both day and night.
Organic Chemistry, a journey divine,
Where possibilities flourish, like stars align.
    Pass this subject with flying hues,
And witness the wonders it will infuse.
For in the realm of atoms and bonds so true,
Organic Chemistry holds a future for you.

ABOUT THE AUTHOR

Walter the Educator is one of the pseudonyms for Walter Anderson. Formally educated in Chemistry, Business, and Education, he is an educator, an author, a diverse entrepreneur, and he is the son of a disabled war veteran. "Walter the Educator" shares his time between educating and creating. He holds interests and owns several creative projects that entertain, enlighten, enhance, and educate, hoping to inspire and motivate you.

Follow, find new works, and stay up to date
with Walter the Educator™
at WaltertheEducator.com

www.ingramcontent.com/pod-product-compliance
Lightning Source LLC
Chambersburg PA
CBHW051343150726
48000CB00003B/1027